Eh, Canada? Clever Creations

Pocket-Sized Canadian Trivia Books

Written and
Illustrated by
Jacqueline Cooper

Published by Little Goodbyes Press

www.littlegoodbyes.ca

Large Print Edition

Set in 16pt Lexend

Publisher's Cataloguing-in-
Publication data

Cooper, Jacqueline

Eh, Canada? Clever Creations /
Jacqueline Cooper.

ISBN 978-1-997874-06-5

1. Canada—Miscellanea.
2. Curiosities and wonders—Canada.
3. Inventions—Canada—Humour.
I. Title.

Eh, Canada?
Clever Creations

Pocket-Sized Canadian
Trivia Books

Jacqueline Cooper

Little Goodbyes Press

Table of Contents

About the Series

Eh, Canada? is a series of tiny books stuffed with big trivia.

Each volume dives into the quirky, peculiar, and sometimes downright ridiculous sides of Canadian life including weird laws, clever creations and tasty foods

to strange animals,
haunted places, and more.

They're pocket-sized, easy
to flip through, and
perfect for stocking
stuffers, coffee tables, or
anywhere you need a dose
of, "Wait, is that really
true?"

Because let's face it: Canada is polite, proud, and just a little bit absurd. And that's exactly why we love it.

Canada's Clever Creations

Canada is famous for hockey, maple syrup, and apologizing too much. But tucked behind all that politeness is a surprising streak of brilliance.

Canadians have been quietly changing the world for over a century and not just with big breakthroughs that everyone knows, like insulin and the Canadarm, but also with little everyday life-savers like the garbage bag and the paint roller.

This book celebrates Canadian creativity in all its forms. Some inventions are medical miracles, others are quirky kitchen classics, and a few will make you wonder how we ever lived without them. Each page gives you a quick peek at an idea that started in Canada and spread far beyond our borders.

So grab a butter tart, settle in, and prepare to be amazed at just how many times Canadians have looked at a problem and said, "We can fix that, eh?"

Medicine & Science

Insulin (1921)

Before 1921, a diagnosis of diabetes was basically a death sentence. Enter Canadian doctors Frederick Banting and Charles Best, who teamed up at the University of Toronto and changed the world.

They figured out how to extract insulin from a dog's pancreas and use it to lower blood sugar in humans.

It worked. Suddenly, people with diabetes had a treatment and a shot at long, healthy lives.

This discovery was so groundbreaking that Banting ended up selling his patent to the university for just one dollar. He didn't want anyone to get rich off it, he wanted insulin to save lives, not line anyone's pockets.

Fun fact
At only 32, Banting remains the youngest Nobel Prize winner in medicine.

Pablum (1930)

In 1930, doctors at Toronto's SickKids Hospital created Pablum, one of the first packaged baby cereals fortified with vitamins and minerals such as Vitamin D and Iron.

The goal was to make it easier for families to boost

nutrition at a time when many children weren't getting enough from their diets due to declining breastfeeding rates and to help protect against rickets, anemia and other common concerns.

The cereal itself was a bland, powdered mix that could be stirred into breastmilk or formula by parents.

It wasn't perfect, and modern feeding advice has moved well past it for the standard first food, but at the time, it was a breakthrough in making fortified foods widely available.

Fun fact
The word "pablum" comes from the Latin for "food."

Thanks to this invention, it also became slang for something dull or bland which might be the most honest product review ever.

Electron Microscope (1938)

Sometimes regular microscopes just aren't enough.

In 1938, James Hillier, a young scientist from Brantford, Ontario, and his colleague Albert Prebus built the first practical transmission electron microscope (TEM) in North America while working at the University of Toronto.

Germany had built an early version in 1931, but Hillier and Prebus were the ones who proved the technology could really work outside a lab prototype.

Instead of light, their microscope used beams of electrons to magnify objects thousands of times more than any light microscope could manage.

Suddenly, scientists could see viruses, fine cell structures, and materials at the atomic level. Without it, much of modern medicine and nanotechnology wouldn't exist.

Fun fact
Hillier and his colleague Albert Prebus were grad students when they pulled it off. Hillier later joked, "That he only took the project because he thought it might get him a job." Instead, it got him a place in science history.

Pacemaker (1951)

Canada isn't just good at keeping hearts warm with maple syrup, we also figured out how to keep them beating.

In 1951, Winnipeg engineer John Hopps built the world's first external artificial pacemaker.

He wasn't even a doctor. Hopps was a radio engineer working with the National Research Council when he discovered that applying electrical pulses could restart a stopped heart.

The machine he built was about the size of a microwave, so you wouldn't exactly strap it on for a jog, but it laid the groundwork for the tiny implantable pacemakers we know today.

Fun fact
In the early 1940s, Hopps wasn't working on hearts at all, he was busy experimenting with how to pasteurize beer using radio and microwave waves. Only in Canada do you go from saving beer to saving lives.

Chirped Pulse Amplification (1980s)

In the 1980s, physicist Donna Strickland, a Canadian graduate student at the time, and her supervisor Gérard Mourou, developed a new way to amplify laser pulses without destroying the laser itself.

The technique, called chirped pulse amplification (CPA), stretches a laser pulse, amplifies it, and then compresses it again to create an incredibly powerful, ultra-short burst of light known as a femtosecond laser.

This breakthrough made it possible to cut and shape materials with extreme precision.

Today, CPA lasers are used in eye surgeries like LASIK, in delicate manufacturing, and in scientific experiments that probe matter at the atomic level.

Fun fact

In 2018, Strickland became the third woman ever to win the Nobel Prize in Physics, and the first in more than half a century, an achievement that put both her and Canada's University of Waterloo firmly in the global spotlight.

Sprinkles (1996)

Decades after Pablum, SickKids Hospital was at it again. In the 1990s, Dr. Stanley Zlotkin developed Sprinkles, tiny sachets of powdered vitamins and minerals that could be mixed into whatever food children were already eating.

Unlike Pablum, Sprinkles weren't a cereal or a meal on their own. They were designed for use in developing countries, where kids often suffer from "hidden hunger" those who did not have enough iron, vitamin A, or zinc in their diets.

The single-serving packets made it cheap and simple for families to boost nutrition without changing their cooking.

Today, Sprinkles are a World Health Organization recommended product, used in more than 60 countries to help fight childhood malnutrition.

Fun fact

The breakthrough wasn't just the powder, it was the packaging.

Zlotkin was eating out one day when he noticed little packets of sugar, salt and pepper. If seasonings could be portioned out that way, why not vitamins?

He realized a single-serving packet would be foolproof, even for parents who couldn't read complex instructions.

Just open and sprinkle.

Every Day Life

Egg Carton (1911)

Ever crack eggs on the way home from the store? In 1911, British Columbia newspaper editor Joseph Coyle overheard a hotel owner and a farmer arguing over who was to blame for broken eggs.

Coyle came up with the egg carton: a cardboard tray with little pockets to keep each egg safe.

Coyle's first egg cartons were hand made but he eventually designed a machine so they could be produced in greater numbers.

He founded the Egg Safety Carton Company who marketed it under the slogan, "It saves you more than its cost".

Fun fact
Coyle's original cartons held 12 eggs, setting the standard "dozen" packaging that's still used around the world today.

Paint Roller (1940)

Before the 1940s, painting a room was a slow, streaky chore with a brush. Then Toronto inventor Norman Breakey came up with a simple but brilliant idea: wrap an absorbent material around a rotating frame, dip it in paint, and roll it smoothly across a wall.

Breakey patented his roller in Canada in 1940 and produced them in a small factory he established in Toronto.

After the war, he sold at least 50,000 under the name Koton Kotor, and Eaton's carried them as the TECO roller. But without international protection, his design was copied abroad and he missed out on the global market his invention created.

Fun fact
Even though Breakey didn't cash in big, his paint roller completely changed how people tackled DIY jobs, turning weekend walls into a Canadian claim to fame.

Garbage Bag (1950)

Before the 1950s, trash was tossed into metal bins or barrels which were messy, smelly, and not fun to clean. Enter Winnipeg inventors Harry Wasylyk and Larry Hansen, who created the first green polyethylene garbage bags in 1950.

Their original idea wasn't for households at all, the bags were designed for hospitals, to keep medical waste contained and sanitary.

Once the idea caught on, Union Carbide licensed it for mass production, and sold it under Glad Garbage Bags, and garbage collection was never the same again.

Fun fact

The very first garbage bags were green, not black. Today, most cities colour-code their bags: black for trash, blue for recycling, green for compost, so you could say Canadians basically invented the colour-coded cleanup.

Electric Wheelchair (1953)

When war veterans returned home with serious injuries in the 1940s, mobility was one of the biggest challenges. Enter Canadian inventor George Klein, an engineer at the National Research Council.

In the early 1950s, he developed the first electric wheelchair, a design that gave paralyzed patients far more independence than manual chairs ever could.

Klein's chair used joystick controls and a battery-powered motor, could be used both indoors and outdoors, and it quickly became the foundation for modern power wheelchairs. His work dramatically improved the quality of life for people with limited mobility, and his design was adopted worldwide.

Fun fact

Klein never patented the wheelchair. He considered it a public service, not a for-profit opportunity, which is why his design was freely shared and copied around the globe.

Wonderbra (1964)

In 1964, Canadian designer Louise Poirier changed lingerie forever. Working at Canadian Lady Corset in Montreal, she engineered the Wonderbra Model 1300, the push-up bra that promised "lift and cleavage like never before."

Poirier's design used a mix of angled padding, elastic, and underwire, and it quickly became an international sensation. Copycats followed, debates raged, and the Wonderbra turned into one of the most famous undergarments of all time.

Fun fact

Poirier's Model 1300 had 54 separate design elements and is almost identical to the ones still made today. This design wasn't just fashion, it was engineering with lace.

Technology & Space

Telephone (1876)

When it comes to claiming the telephone, both Canada and the United States say "Hello" first. Inventor Alexander Graham Bell was Scottish-born, lived in both countries, and filed his famous patent in the U.S. in 1876. However, much of the work that led up to it happened in Brantford, Ontario, where he had his family home and workshop.

In the summer of 1876, Bell transmitted a voice signal over a telegraph wire from his Brantford study to a nearby village, marking one of the earliest successful long-distance calls.

The patent went south, but Canada still proudly calls Brantford "The Telephone City."

Fun fact
Canada and the U.S. still quietly compete over who invented the phone, but even Bell himself called Brantford "the birthplace of the telephone." So maybe the last word should be his, eh?

IMAX (1967)

Canada's love of big ideas got literal in 1967. A group of filmmakers and engineers (Graeme Ferguson, Robert Kerr, William Shaw and Roman Kroitor) wanted to create a movie experience that filled your entire field of vision.

Their new projection system, first shown at Expo 67 in Montreal, used massive reels of 70mm film and a curved screen to pull audiences right into the picture.

The name came from "Image Maximum," and the impact lived up to it. IMAX theatres soon appeared around the world, turning ordinary movie nights into full-body experiences.

Fun fact
The first permanent IMAX theatre opened in Cinesphere in Toronto in 1971. It's still standing today at Ontario Place, currently closed for redevelopment but should be soon ready for its next close-up.

Canadarm (1981)

In the 1980s, Canada quite literally reached for the stars. A team of engineers at Spar Aerospace, working with the National Research Council and NASA, built the first Canadarm which was a robotic arm designed to lift satellites, move cargo, and even grab spacecraft in orbit.

It flew for the first time aboard the Space Shuttle Columbia in 1981 and quickly became a symbol of Canadian engineering around the world. Controlled from inside the shuttle, the arm moved with incredible precision, earning the nickname "the space crane."

Fun fact
In 2001, the Canadarm1 performed the first robotic handshake in space with Canadarm2.

The original Canadarm retired after 90 missions and now lives at the Canada Aviation and Space Museum in Ottawa, so you can grab a selfie with it anytime.

Its successor, Canadarm2, is still working on the ISS today, carrying on the tradition.

A third generation, Canadarm3, is already in the works for Gateway (a lunar outpost that will enable sustainable human exploration of the Moon) and is proof that Canada's reach just keeps extending.

Braille Printer (Converto-Braille) 1972

Quebec inventor Roland Galarneau, a largely self-taught machinist who was legally blind, built the Converto-Braille, a computerized machine that translated text into Braille at roughly 100 words per minute.

It linked an electromechanical computer to a teletype/embosser so printed text could be converted quickly into Braille pages.

The device helped speed up production of Braille textbooks; by the mid-1970s his organization was supplying materials across Quebec.

Today, an original machine is displayed by Ingenium at the Canada Science and Technology Museum in Ottawa.

Fun fact
Galarneau started the project so people who couldn't read Braille themselves could still produce materials using Braille bringing

newspapers and textbooks within reach much faster than manual methods.

Artificial Intelligence (2018)

Before computers could write essays, chat, or spot faces in photos, they had to learn to learn. That breakthrough came from a group of researchers who taught machines to recognize patterns the way human brains do.

Two of them, Geoffrey Hinton at the University of Toronto and Yoshua Bengio at the Université de Montréal, helped pioneer the field of deep learning, a branch of artificial intelligence that uses layers of simulated neurons to process information. Their discoveries powered everything from voice assistants to medical imaging and modern AI systems.

Fun fact
In 2018, Hinton, Bengio, and their collaborator Yann LeCun received the Turing Award, often called the "Nobel Prize of Computing." Canada can proudly claim two of the three "Godfathers of Deep Learning."

Defence & Survival

Gas Mask (1915)

When poison gas
appeared on the
battlefields of the First
World War, soldiers had
almost no protection and
many resorted to holding
wet cloths or even rags
soaked in urine over their
faces to survive. A
Newfoundland doctor

named Cluny MacPherson decided to fix that.

In 1915, he designed a fabric hood soaked in chemicals that neutralized chlorine gas. It covered the entire head with a glass eyepiece and a breathing tube, offering far better protection than anything used before. The British Army quickly adopted his design, calling it the "British Small Box Respirator," and it became the model for modern gas masks.

Fun fact
MacPherson tested early versions on himself by breathing in chlorine gas in his lab to make sure the mask worked. Once, when he was in the hospital after a test that hadn't gone well, he noted, "I

took this Viyella and mica out of my pocket and got a sheet of paper and cut out the design of the helmet and got the nurse to sew it up for me and put it back in my pocket." [sic] From a steady nurse's stitches to the battlefield, it proved that Canadian ingenuity can come from anywhere.

ASDIC (SONAR)
1917

Long before radar guided planes through the sky, scientists were trying to use sound to detect submarines under the sea.

Working with the British Admiralty, Canadian physicists Robert William Boyle and British physicist

Albert B. Wood were both instrumental in developing ASDIC, the precursor to modern sonar. This was used as an early underwater detection system that sent out sound waves and measured their echoes to locate and track submarines beneath the surface.

During the First World War, the technology was still experimental, and only a few dozen ships carried prototype sets by 1918. It wasn't until the late 1930s and 1940s that ASDIC became standard equipment across Allied fleets and was installed on thousands of ships.

Fun Fact

In 1939, in response to a query from the Oxford English Dictionary, the Admiralty made up the story that the letters stood for 'Allied Submarine Detection Investigation Committee', and this is still widely believed, though no committee bearing this name has ever been found.

Walkie-Talkie (1937)

When you think of the walkie-talkie, you probably picture Allied soldiers barking orders through crackly radios. The first truly portable version came from Donald L. Hings, a Canadian engineer and inventor from Alberta.

In 1937, Hings built a small, battery-powered AM transceiver for his mining company, to help crews stay in contact in remote areas.

When the Second World War began, the Canadian government adapted his design for military use, producing thousands for Allied troops.

His compact radio became known as the "walkie-talkie."

Fun fact

Hings originally called his invention the "packset." The name "walkie-talkie" caught on after soldiers started using it in the field in WWII and it's stuck through to today.

Snowmobile (1937)

Canadian winters can make travel almost impossible, but that never stopped Joseph-Armand Bombardier. In 1937, the young mechanic from Valcourt, Quebec, built the first practical snowmobile. It was a lightweight, tracked vehicle that could glide over deep snow.

Bombardier originally designed it so rural doctors and mail carriers could reach remote towns in winter. His early models, called B7 Snowmobiles, could carry up to seven passengers and were steered with skis up front. During the Second World War, the Canadian military ordered versions for rescue and transport in snowy terrain that could carry up to 12 people.

Fun fact
Bombardier's company later became a global transportation giant. The humble snowmobile that started as a hometown project eventually led to trains, planes, and even jets with his name on them.

G-Suit (1941)

During the Second World War, fighter pilots were pushing aircraft to new limits and sometimes blacking out from the extreme forces of sharp turns and dives. At the University of Toronto, Dr. Wilbur R. Franks came up with a way to stop it.

He created the world's first anti-gravity suit, or G-suit, by filling rubberized tubes with water and later air pressure. When pilots pulled high Gs, the suit squeezed their legs and abdomen to keep blood from draining away from the brain.

The design was quickly adopted by Allied air forces and became the basis for modern G-suits that are worn today by pilots and astronauts.

Fun fact
The suit's name during development was Franks Flying Suit.

Food & Drinks

Peanut Butter (1884)

Before lunchboxes made it famous, peanut butter began as peanut candy. In 1884, Marcellus Gilmore Edson of Montreal patented a way to mill roasted peanuts between heated surfaces to create a smooth paste that could harden as it cooled. He called it peanut candy and suggested mixing in sugar for taste and texture.

Edson's patent was the first to describe this peanut paste and the process behind it, making it the foundation of the modern peanut butter industry that is going strong today.

Fun fact
His patent suggested
using a grain/flour grinder
to make the paste, a
simple start for one of the
world's favourite comfort
foods.

Butter Tarts (early 1900s)

Sweet, sticky, and proudly Canadian, the butter tart first appeared in community cookbooks in Ontario in the early 1900s. The earliest known published recipe comes from the Women's Auxiliary of the Royal

Victoria Hospital in Barrie, Ontario printed in 1900.

Made with butter, sugar, syrup, and eggs baked in a flaky pastry shell, butter tarts became a staple of church fundraisers and kitchen tables across the country. Families quickly began arguing over one essential question: raisins or no raisins?

Fun fact

The butter tart is now so beloved, that Ontario towns like Midland host an annual butter tart festival, celebrating Canada's stickiest culinary debate, and where over 300,000 butter tarts are sold to over 60,000 visitors in only one day.

Nanaimo Bars (1950s)

Rich, layered, and no baking required, the Nanaimo bar is a West Coast classic. The earliest printed recipe appeared in the 1950s in a Nanaimo, British Columbia, cookbook, though similar "chocolate slices" were known across Canada.

A true Nanaimo bar has three layers: a crumb-and-coconut base, a creamy custard filling, and a smooth chocolate top. It's now a fixture at bake sales, potlucks, and coffee shops from coast to coast.

Fun fact
The city of Nanaimo, B.C., even has a self-guided "Nanaimo Bar Trail,"

featuring everything from traditional squares to ice cream and martinis inspired by the dessert. There's even a Nanaimo Bar pedicure to help wind down after visiting the over 40 locations on the map.

Poutine (1950s)

Few dishes spark as much Canadian pride, or as many arguments about who invented it, as poutine.

In the late 1950s, diners across rural Quebec started mixing French fries, cheese curds, and hot gravy into one gloriously messy meal.

Most stories trace it back to Warwick, Quebec, where restaurant owner Fernand Lachance is said to have first tossed the cheese curds onto fries at a customer's request. When he saw the gooey result, he reportedly exclaimed, "Ça va faire une maudite poutine!" or roughly, "That's going to make a darned mess!"

Later adding gravy to keep it warm.

But others say that Drummondville's Jean-Paul Roy is the true inventor since Le Roy Jucep was the first to sell poutine with three combined ingredients, and holds the copyright registration certificate, issued by the Canadian

Intellectual Property
Office.

Fun fact
Poutine didn't become
popular across Canada
until the 1980s, when fast-
food chains added it to
their menus. What began
as a happy accident in
rural Quebec is now an
international comfort food
classic.

Hawaiian Pizza (1962)

In 1962, a Greek-Canadian restaurant owner named Sam Panopoulos decided to have a little fun with his pizza toppings at the Satellite Restaurant in Chatham, Ontario.

He opened a can of pineapple and tossed some pieces on a ham pizza just to see what would happen.

To his surprise, customers loved the sweet-and-salty mix. He named it after the brand of canned pineapple he used, Hawaiian, and one of the world's most debated pizzas was born.

Fun fact

Despite the tropical name, the Hawaiian pizza was 100% Canadian. Even the controversy that followed, whether pineapple belongs on pizza, started right here.

Canola Oil (1970s)

In the 1970s, Canadian agricultural scientists Baldur R. Stefansson and Keith Downey developed a new variety of rapeseed with lower levels of bitter compounds, making its oil safe and pleasant to eat. They called it canola.

Canola quickly became one of Canada's most important crops and is now a staple cooking oil around the world. It's also used in everything from salad dressings to biodiesel.

Fun fact
Canola is actually an acronym for "Canadian oil, low acid".

Thanks for Coming Along

Thanks for coming along on this tour of Canadian creativity.

From frozen frontiers to kitchen counters, Canadians have been quietly inventing, improving, and experimenting their way through history.

Some ideas changed the world; others just made life a little easier (or tastier).

These pages are proof that inspiration can strike anywhere, whether it's a lab, a workshop, or a small-town diner.

About the Author

Jacqueline Cooper is a Canadian writer with a sharp eye for the offbeat details hiding in everyday life. She also illustrated this book, keeping the art as Canadian as the creations inside.

When she's not digging through articles for oddball trivia, she runs Little Goodbyes Press, a small indie Canadian press creating books that mix humour, heart, and a dash of curiosity.

About the Publisher

Little Goodbyes Press is an independent Canadian publisher creating books that celebrate curious details, gentle humour, and unexpected stories.

From children's picture books to quirky trivia, every title is made with care and just enough Canadian flavour to make you smile.

Looking for more of our books?

Be sure to visit us at
www.littlegoodbyes.ca

www.ingramcontent.com/pod-product-compliance
Lightning Source LLC
Chambersburg PA
CBHW020743130626
46554CB00006B/2127